AF679207

AN AG MAN'S TRAVELS

Authored by Fred W. Chase

Compiled by Charles G. Chase

Infusionmedia
Lincoln, Nebraska

Infusionmedia
Lincoln, Nebraska
infusion.media

Printed in the United States

10 9 8 7 6 5 4 3 2 1
First Edition

ISBN: 978-1-945834-35-6

Library of Congress Control Number: [PENDING]

This volume comprises Fred W. Chase's *A Brief Historical Sketch of the Life of Fred W. Chase* (1925), newly typeset, and his travelogue of photographs and notes (1922), reproduced in facsimile. Original spelling, punctuation, and usage have been preserved as faithfully as possible.

This book is set in GT Super, designed by Noël Leu with Mirco Schiavone and Reto Moser, released by the Swiss type foundry Grilli Type in 2018, and inspired by lettering in a newspaper advertisement for Buick automobiles.

TABLE OF CONTENTS

INTRODUCTION

Welcome to the life of one of my patriarchs, Fred W. Chase.

I never knew Fred W. Chase, nor did I know his son, my grandfather, Glen Harold Chase Sr. Yet my great-grandfather had a unique influence on me while I was growing up. We celebrated the birthdays of all of us children, but mine was different. On my birthday, we would all go to Pawnee City, Nebraska. The whole town would show up and throw me a parade. They would follow us out to the cemetery and visit Fred W.'s grave. The mayor would speak, the National Guard would fire their weapons, and then our family would go to my grandmother's house for fried chicken. I don't know how old I was when they finally let me know that I had been born on Memorial Day. Until then, I thought I was pretty special.

I had read *A Brief Historical Sketch of the Life of Fred W. Chase* and had provided copies for my own children. Yet when Don Chase lent me Fred's travelogues, I was mesmerized. I owe Don a great debt of thanks for lending them to me and granting me permission to publish them. I believe this book will be of interest to anyone with enough imagination to travel back in time to the backroads and hamlets of North America in a 1920s Buick five-passenger touring car, when an automobile road trip was still something new and adventurous. Yet it is of even greater interest to me, to my children, and to all the other descendants of this man of agriculture. And it is to them that this book is dedicated.

Speaking of agriculture, in Nebraska, they say, "You know you're a farmer if you have ever driven off the road while looking at your neighbor's crops." That is something that stood out to me as I read these travelogues. There are photographs of tourist attractions and historical sites, but it seems that what Fred most wanted to photograph was farming. How do they harvest bananas? How do they farm in Vermont? Look—a tractor! When he visited universities, he photographed the agriculture buildings, not so much the sports fields.

And while we are discussing agriculture buildings, please take note of the Nebraska Agricultural Engineering Building on East Campus in Lincoln, Nebraska. Fred's son, Leon W. Chase, was not only an instructor there but also founded the world-famous Nebraska Tractor Test Laboratory. What Fred W. did not know when he took the picture was that in 1982 the building would be renamed L.W. Chase Hall, the name by which it is still known today.

As you read about the life and times of this man of agriculture, I hope you, too, will be transported to a period when farming everywhere was changing and becoming a science, with new crops, new methods, and new machinery. It was a pivotal time of change in travel, education, farming, rural America, and, in many ways, the world.

I resisted the urge to edit out the photographs of billboards, roadside clutter, or other seeming nonsense. Leaving it all in adds to the feeling that you are there, seeing these things for the first time. In this book, you get to see it when it was new, through the eyes of a man who knew the past well yet could still dream of the future.

Enjoy.

Chuck Chase

Brief Historical Sketch

of the

Life of Fred W. Chase

Pawnee City, Nebraska

JANUARY FIRST

NINETEEN HUNDRED TWENTY-FIVE

A Brief Historical Sketch

of the Life of

Fred W. Chase

Pawnee City, Nebraska

January 1, 1925

A BRIEF HISTORICAL SKETCH OF THE LIFE OF FRED W. CHASE

Fred Wilson Chase, of the eighth generation since Aquilla Chase settled in America, was a son of Josiah Briggs Chase and Elsie Polina Warren-Chase, was born June 16, 1853. Josiah Briggs Chase, of the seventh generation, was a son of Moses Chase. Moses Chase, sixth generation, was a son of Isaac Chase, of the fifth generation, who in turn, was a son of David Prince Chase, a soldier of the Revolution, of the fourth generation. David Chase was the youngest son of Benoni Chase, of the third generation, who was the son of Moses Chase, second generation, the son of Aquilla Chase, who was born in England in 1618.

I was born in Whitingham, Windham County, Vermont, June 16, 1853. My parents were both from a long list of sturdy New England stock. My father's (Josiah Briggs Chase) genealogical line traces direct to Aquilla Chase, who came across the Atlantic about the year 1636 and settled in the extreme south-eastern part of New Hampshire. The records of the Chase family show that I am a direct lineal descendent of Aquilla Chase, who came to America in 1636, and settled at Hampton, Massachusetts. I think the above to be correct, although my father, Josiah Briggs Chase, said Aquilla Chase settled at Cornish, New Hampshire, but so far as I have found, all the records show that Aquilla Chase was born in England in 1618 and settled at Hampton, Mass., about the year he came across.

My mother, Elsie Polina Warren, I have been unable to find any record to show when or where the first family of Warrens came from to this country. I used to be told in early childhood that my mother was a direct lineal descendent of the General Warren, who fell in the battle of Bunker Hill. In trying to trace this line of descent from General Warren I could not do so, for the records of my great-grandfather, James Warren, seem to have been lost. Still I think there must be more or less truth in the statement that I am a direct descendent of General Warren, for while visiting Bunker Hill Monument a few years ago my attention was drawn to the life-size statue of General Warren, that stands in the rotunda of Bunker Hill Monument, and there being such marked resemblance of features and life size of some of the present generation of the Warren's descendents of this said to be a perfect likeness of General Warren, that if one can depend upon these features being handed down to posterity through generations, then surely my mother was a descendent from this great man, General Warren, although I cannot find any records to prove it.

My first attendance at school was when I was three years old. My name was not put on the school register, for I was not a very regular attendant. Why it was that my mother sent me to school at that age I never knew, but I have always supposed it was because I was so mean that I made life miserable for her, and she thought she could shove me off on the school teacher for a few hours and get a little rest. My mother at that time did nearly all of the spinning, weaving, knitting and sewing of the family apparel. However, but very few months of my life was spent in the schoolroom. Sixteen months would cover it all.

My father owned and operated a Green Mountain farm, and tried to do nearly all the work himself. Lumber and wood had to be got out in winter. The logs had to be hauled to the water-power saw-mill, as there was very little steam in those days and electrical power was unheard of at that time.

Shingles were split and shaved by hand for all the building, and many was the evening that I can remember of my father working clear into the night making shingles, and his only light was from the fire in the fireplace.

I was quite a mechanic as a boy, and used to spend quite a little time during winters in the little improvised work-shop in the wood-house when the weather was moderate enough so I could work in the shop comfortably. About all the evening entertainment I got in those days was what I could get calling at some of the neighbors. One winter I remember very distinctly of going to my father's lumber pile and selecting some of the best boards and taking them to the little shop, hand-planing and making a table for parlor croquet. This venture, although a lot of hard work was connected with its structure, proved a great drawing card for the young people of the neighborhood, and scarcely an evening passed without neighbors gathering for an evening game. The only light we had to play with was a tallow candle my mother had made for use about the house. My mother had what was called in those days candle sticks for the tallow candles. They each had a lip or hook on the side to hang them up with, and I fixed it so we could hang the candle stick up, and I assure you we boys and girls had as much pleasure and sport playing parlor croquet as the young people have nowadays playing games under the most expensive electrical fixtures. At that time tallow candles and light from the fireplace furnished all the light we could have, and we were content to get along with it.

Agricultural methods were very primitive in those days. My mother used to make some butter, and oh! such times she had. The milk was strained into shallow tin pans and set up on a high shelf in the kitchen, where the superfluous dirt collected on the cream, to be skimmed off with the cream in the morning and churned into butter. The churning—that was my job. That old dash churn; I can see it now. There never was but one job in my life that I hated as badly as to manipulate that old dash churn of my mother's, and that was to herd cattle on the prairie. I declared many a time while herding stock that if I had to have such a job for a livelihood I would steal a horse so that I would be put in the pen. No herding cattle or churning with a dash churn in mine.

So the years rolled along, and I was seventeen. School days were over for me, and I thought I would like to go out in company. Fourth of July was approaching with a celebration to be pulled off in a neighboring town, and I wanted to take a good-looking girl with me to the celebration. I had my girl selected, but how, when and where was I to ask her for her company to the celebration. Should I go to her home, go to the front door and rap, and run the risk of having to undergo the embarrassment of asking if I could see their daughter? But fortune seemed to favor me, for I met her one day, and right then and there I said, “Can I have your company to the Fourth of July celebration?” The reply was favorable, and I was all O. K. But my troubles were not over. I had a girl engaged to go to the celebration with me, and as yet no known way of taking her, and it was too far to go afoot, so something had to be done. It seemed I had tackled the easy job first, for it seemed easier to ask the girl to go with me than to ask my father for the old gray mare to take the girl to the picnic. However, I made up my mind that “nothing ventured, nothing won,” so just as I was starting to the field to work one morning I asked father if I could take the old gray mare to carry a girl to the picnic the Fourth, and he said yes, and it would now seem for the most part things for the picnic were nearly completed. Would it be a pleasant or a rainy day? I could not go in the rain, and if it did prove rainy all my preparations would be lost; but the only thing to do was to wait and see. The day dawned pleasant, all went to the picnic and I had a good time, and always thought the girl did. I got her home in decent time, and got home myself not very much worse for the venture.

I was always very much interested in my father’s farm and in helping him conduct little experiments in agriculture. I remember one time my father arranging his business and chores so he could leave home and go down in Franklin County, Massachusetts, to purchase some heifers and cows. He thought he had saved up money so that he could buy four, but when he got there the quality he wanted would cost much more than he had expected, so he only bought two. They were of the Shorthorn breed, and I took great delight in milking and taking care of those two cows. One of these was a hard milker, and I never liked this one as well as the other. In those days there was no known way of testing milk as to its richness, except to set it in tin pans and look for the thickness of the cream, but with the assistance of my mother I was able to go one better. She had some clear glass tumblers that she had bought of a tin peddler by saving up some rags and trading for these tumblers. I can see now how proud she felt the first time she set her table with those new tumblers. My mother selected two of these tumblers that were of the clear glass, and that night I milked each cow separately and brought the milk to the house and strained some of each cow’s milk in one of these glasses, just about the depth of the milk in the tin pans, and set it away to raise with the other milk. As the glass of these tumblers was quite clear we could easily tell which cow produced the thickest cream. While the method was very crude as compared with our present day Babcock test, yet it served the purpose much better than any method heretofore found, and I felt that we had advanced at least one step toward testing the value of a cow over anything we had previously known that would give any idea of how rich milk our cows were producing. And to this day, if I did not have a Babcock tester to test the milk of our cows, the tumblers would give us a much better idea of what our cows were doing than merely guess work.

I was always very much interested in dairying. I remember very distinctly while in the dairy exchange work of making a remark that I had been in the dairy business all my life. A quick-witted Irishman said: "You say you have been in the dairy business all your life?" "Yes, sir, I have," I replied. "Then tell us, will you, what did you do the first year?" My reply was, "I milked, if you please."

There was no such a thing as a dairy school when I was a boy. If we learned anything we had to work it out ourselves without an instructor. We had no silos. In those days we had June grass and Herd grass, dead ripe. Herd grass is just about as good for milch cows as the common bunch grass that grows in our Nebraska timber.

I was married on Christmas eve, December 24, 1874. It is needless for me to say anything about my courtship days, for I married my first and only love, and she did her share of the courting. We commenced housekeeping the following spring. Our stock in trade at that time consisted of one work horse, two cows, and a pair of yearling steers and five hens. We had no farming implements whatever, as the farm we settled on adjoined my father's farm, and I got along quite well by borrowing tools from him. My father gave me one of the yearling steers, and to have a pair of oxen coming on I must procure a mate for this, as nearly all the farming work in those days was done with oxen. It was not only necessary, but very essential, that I have a pair of oxen coming on. I remember very distinctly my trip to find a mate for this steer. I found one that looked very good to me, and bought it and started home with it at the end of a rope. I had gone nearly a mile when I turned to see how my purchase looked. At the first glance I was sick, for there stood a hollow backed, coarse haired critter. How did I get so terribly stung? I looked my purchase over right there on the road all alone, and found the steer had some good points: a beautiful face and countenance, a good thick hind quarter, a good straight hind leg, a clear eye and a clean muzzle. The man I bought him of was an expert in the selling game. He had drawn my attention to the beautiful face, heavy quarter, the straight leg, and I looked so much at these that I did not see the poor points in that steer until I was all alone. Then I could see them very plainly, and I made up my mind right there, and I know where that spot is and the birch tree that I stood under, that I would never let another man detract my attention from the poor points of a steer to the good ones to such an extent that I could not see the poor ones. This was a good lesson I bought, and I will never forget it.

My farming operations in Vermont consisted chiefly of making maple sugar in the spring, cutting and hauling wood and logs in the winter—a mighty lot of hard work and not much pay. Cleaning the land of rocks so that the grass for hay could be cut with a mowing machine. When I commenced farming for myself there were very few mowing machines, and the price of one was such that I could not own one, so I had to cut all my grass for hay with a scythe, rake it with a hand rake, and pitch it by hand. I was very anxious to get the land smooth, so that as soon as I could get the machinery I would have the land ready for its use.

It was quite a job for the common farmer in those days. Grain farming was very limited. If a farmer raised one hundred bushel baskets of ear corn he was some farmer, and it was my desire and aim to raise that much in one season.

Mating and matching steers to grow up into well trained oxen was quite an industry, and I soon learned that a little grain mixed in the feed made a great difference in the looks of the steer, he showed up so much better to the buyer when his hair laid smooth and sleek, and I spent much time in brushing and fixing them. The town people used to have what they called a cattle fair every fall, and the judges would issue what they called honorary premiums, and the farmers took great interest and had a lot of amusement in taking their products to the fair and in having them judged. I remember one fall of taking the premium over all others for trained steers. This pair of steers were one year old past; I drove into the crowd of people without a yoke or halter, and with my whip could drive either tandem or abreast.

Farming in those days in Vermont was a slow, up-hill business, and I commenced to think of another location. I remember about this time of fattening four shoats for market, hoping I could make a dollar or two. When I got them as I thought fat enough for market, I commenced to look about for a market. I visited the butcher shops in four near-by towns, and none of them would take these shoats at any price. Their excuse was that western pork was much better. There I was with a herd of four fattened shoats on my hands, weight about 200 pounds each, and no market. Finally, by perseverance, I found a man in a country town, away from the railroad, where some extra building was in progress, who said he would take one a week if I would dress and bring them to him. There was no other way out of the dilemma. I dressed one a day and took the next day to deliver it. There were no autos in those days. Everything went well until the last one, and then my market dropped. When I got there the man was off—too much meat on hand—said he wished I had brought but half of it. I asked him what he thought I could do with the other half; keep it and fatten it a while longer? With some coaxing he took it all. This was my first and last venture in the feeding business in Vermont. This taught me the lesson if I was to make a living at farming I must go where there was a market for my product. These things continued for a year or more. Whenever the opportunity opened up I would talk about the west, and read about it. Some friends went to Iowa and I wrote to them, and they told me there was always a market for farm products every business day in the year, and all the farmer had to do was to load up his produce and go to town that day, and get the cash at whatever the market value was. This looked good to me, and that a man and team could plow two acres a day and not hit a rock. This, too, looked good to me. My last job with the plow in Vermont was with a pair of oxen one Saturday forenoon. I worked all forenoon and plowed seven rounds fourteen rods long, less than one-eighth acre, and a man dug the stone out of the furrow as I went along.

That day while cleaning up for the noon-day meal I said to my wife, "I have done my last job of plowing on this farm. If there are places out west where a man can plow two acres a day and not hit a stone, I would like to try it." So the next spring, which was the spring of 1885, found myself and family headed for the state of Nebraska—some jump for the first one—and we got here and settled on one of the poorest farms in eastern Nebraska. When I saw what I was up against I named this farm "Hard Scrabble Ranch," a title that sticks to it to this day. I could plow two acres a day and did not have to pay out a lot of money each year for commercial fertilizer. This was somewhat different from conditions on a Vermont farm, and if I had anything to sell I could get cash for it according to its value any business day of the year.

Our First Nebraska Home Was a Dug-Out.

This story would not be complete unless I told my first experience handling barnyard manure in Nebraska. This ranch I moved onto had on it a typical western stable (one of those that a person has either to haul away the manure or move the stable) so I started hauling the manure on the poor spots of soil, spreading at the rate of forty loads to the acre, as had been the custom in Vermont. A neighbor, passing that way, said, "Chase, you will not raise a thing if you manure it that way." I thought I knew as much about manure as he did, so I went ahead and moved that hill of manure, so that I need not move the stable, and spread it on at the rate of forty loads to the acre. The result was not a nubbin was raised on the patch the first year, but for each succeeding year for seven years a good crop of corn matured. After that I do not know what it did, for I only lived on "Hard Scrabble Ranch" eight years. After my first year's experience with manure in Nebraska I used only about eight loads to the acre, and always got good results from its use applied at this rate.

When I left "Hard Scrabble Ranch" I moved to Norton County, Kansas. Was there about eighteen months, putting in a crop the spring I arrived and raising nothing. I wintered over and put in another crop, and the following August the hot winds burned it all up. These two years were the worst I ever experienced in farming. They were much worse than farming on "Hard Scrabble Ranch," in Pawnee County, Nebraska, so I reversed the rule of etiquette established by early pioneers, never to turn your back to the setting sun, but to face it. I turned my back to the setting sun and came back to Pawnee County, Nebraska, and rented land. There was no trouble in renting land at that time (could have rented one thousand acres within twenty-four hours), and gave as rent one-third of the crop, the owner not charging me anything for the pasture and hay land. I lived on one farm five years, and saved enough to make the first payment of $1,000 on what is known over eastern Nebraska as the Chase Farm, and had stock and tools enough to equip the farm. The purchase of this farm was made in December, 1899, and as my lease for the farm I was living on did not expire until the spring of 1901, I rented the farm and did not move on the so called Chase Farm until the last days of February, 1901. Then commenced an up-hill climb. While the most of this farm laid well, yet many acres were completely worn away. In some places there was not soil enough to raise a cockle burr, and the remainder of it needed a dose of good farming. Alfalfa was at about that time commencing to show its worth, not only as a hog producer but as hay for dairy cows and growing stock. I reserved from the tenant about five acres where I wanted hog pasture.

The Home We Erected on The Chase Farm

I fitted the land the best I knew and planted it to alfalfa in the early spring. The seed came good, and with it millions of weeds. I let it all grow until a foot high, and then mowed it down and left it on the land, the result being that the alfalfa was nearly all ruined. I ought to have known better. No man with any sense at all would have performed such a silly operation and ever expect to get anything. My idea was to get rid of the weeds, thinking the alfalfa would come up from the roots, similar to red clover and timothy. It did not, and I had to learn by experience, the best teacher yet. But western Kansas was producing acres of alfalfa, and Professor Coburn, of Kansas, said it would grow anywhere, so I was not discouraged, and started new again, and tried it with the barnyard fertilizer, and soon learned that the fertilizer was a great help in producing good strong alfalfa, and it was not long until I had forty acres of as fine alfalfa as ever grew in Nebraska.

I learned by experience that ten pounds of good, clean alfalfa seed is plenty of seed per acre to produce a good stand of alfalfa. The best asset to the Nebraska farm that I know of is alfalfa. The more I studied the experiments conducted by the Nebraska Experiment Station the more I was convinced that great advantages for Nebraska farmers were being worked, and that more of them should take advantage of these experiments.

In reference to my work and experience as a promoter of better agriculture, I really don't know what to say. One thing I think will not come amiss, and that is I learned that the nearer I could get to the farmers the more good was done. I found it much better to recommend such things in agricultural work as would come within the means and ability of the common farmers, rather than to advocate something away beyond or above his comprehension. It was always my desire while I was in Farmers' Institute work to visit as many farms as I could at the different towns in which I was billed to deliver an address at the Institute, my reason for so doing being to get better acquainted with the farmers' ideas of farming, and if I could see anything that would prove an improvement or be of advantage over his way of doing, to try in some way to encourage him to take a method that would increase his crop yield and reduce the cost of production. I did not keep an exact record of the different farm homes or towns that I visited, but this I have learned as the years have rolled along, that there are but few towns or farms where I visited that my talks, or at least some parts of these talks, did not leave in the minds of these farmers ideas that stand as living monuments of progressive agriculture, and great advancement made and encouraged, and the great and good work that was recommended in farmers' institutes is still living and growing.

Before I commenced in the Farmers' Institute work I had been experimenting with different varieties of corn, and had done considerable dairy work. The result of most of these experiments I had reported to the State Agricultural College. My work in this line proved to be very interesting, and attracted quite a little attention not only to the college, but among some of the best farmers of the state.

I have been identified in some way with dairying nearly all my life. My first attempt in Nebraska was with the Shorthorn breed of cattle, and I experimented for several years with them. I found this breed to be a failure as a profitable dairy breed, and I learned that if one was to succeed in any branch of agriculture, a breed that for years has been bred for a purpose must be used to make a financial success, or, in other words, if one wants to feed cattle for beef, use the beef breeds; if for milk production, use the milk breeds. There is no such thing as a profitable dual purpose breed—they are not constituted both for profitable beef and milk production.

I gathered records of production from state colleges of both the Holstein-Friesian and Jerseys, and found that the Holstein led in both milk and butter production, and chose the Holstein for our choice of dairy breeds on the Chase Farm. Both of these breeds have centuries of scientific dairy breeding back of the present day stock, and both are profitable dairy breeds in their environments. My sons, who were with me on the Chase Farm, were very much interested in dairying; they all seemed to be natural born dairymen, and took so much interest in feeding and milking that I turned almost the entire management of the herd over to them, with occasional consultation. As soon as the boys assumed the management they adopted a system of testing each individual cow, and soon as an unprofitable cow showed up in the test she was eliminated from the herd, and it was not long until we had a herd of cows, every one of which showed a profit in production, and the boys took great delight in the care of this herd, and in showing it to visitors that came to visit the farm from all parts of the central west, until the herd was disposed of at the time the Chase Farm was sold, November 13th, 1919, when both the farm and the dairy herd, as well as the other livestock and farm equipment were sold at auction. Casey Turnbull, of Clay precinct, bid off the farm at $233.25 an acre.

While I was experimenting with corn, I was also using what was known at that time as the "King Road Drag," and kept and made for the college a very accurate report of its use and the results of its continued use on the highway. This report seemed to and did attract attention among all classes of people in Nebraska, as these reports of the road drag showed that the right use of the road drag was the most economical method of highway improvement that as yet had ever been shown to Nebraska road supervisors.

My reports of corn improvement and the use of the road drag attracted so much attention that I was asked by Mr. Snyder, who was then at the head of the Farmers' Institute department, if I could not make these reports in public meetings. I told Mr. Snyder I could try, and if I did not make a go of it I could go home. My first attempt was at a farmers' institute in Gage County, Nebraska, at a little country church, east of Beatrice.

My address here was very plain and crude, yet the listeners were very much interested and gave me quite a send off. While this happened in the early commencement of state Farmers' Institute work, I am often reminded by some who attended this meeting that they still cling to some of the ideas that were brought out at that time.

In my opinion Farmers' Institutes. through the efforts of the State Agricultural College, have done more to improve agricultural methods than any other one feature I know of.

Since we sold the farm Mrs. Chase and I have been on the retired list. In the meantime we have taken a sight-seeing trip, having traveled twenty thousand miles or more, over seventeen thousand of it by auto. During the trip we were in Old Mexico and Canada, and visited and passed through thirty-eight states, and crossed the continent in an auto. After our return home I was invited to make a public account or description of our trip to the student body of the Agricultural School during the winter agricultural meetings, the first week in January, 1922, at the State Farm in Lincoln. The exact date of the month I cannot at this time recall.

I went to Lincoln April 14th, 1922, to receive a certificate of honor that had been issued to me by the faculty and regents of the State University for the promotion of better agriculture and good roads over the state. I never was a politician of any kind, but when I was presented with this certificate by the hand of the Chancellor of the University, Samuel Avery, at the Temple Theatre in the presence of Dean E. A. Burnett and other officers of the University, and a large assembly of people from over the state, it was the proudest day of my life, and I consider that I had done more good and had built greater monuments in agricultural improvement for the state than had I been in politics at the state capitol. The commencement address was made by the Hon. M. L. Corey, attorney for the Federal Land Bank of Omaha.

I am a member of the Lincoln Chapter of the Nebraska Society of the Sons of the American Revolution, and have been a member of the Chapter ever since its organization. My ancestry in the Revolution consists of three soldiers: 1st, David Prince Chase, private, Col. Leonard's regiment, Massachusetts; 2d, Samuel Parker, private, Col. Docie's regiment, Massachusetts, who was also one of George Washington's body guard for four years; 3d, Jesse Briggs, private, Col. Bradford's regiment, Massachusetts. My children are all eligible to membership in the S. A. R. and D. A. R.

October 20, 1924, my family consists of my wife and nine children: seven boys and two girls. All but one boy, Floyd Josiah, have homes of their own.

My wife, Anna Murdock Chase, was born in Whitingham, Vermont, June 9th, 1856. Her parents, Charles Parker Murdock, her father, and Abigail Wheeler Atherton, her mother, were of the sturdy New England stock. She commenced to teach in the country school when only thirteen years old, and taught continuously for five years, or until we were married, December 24th, 1874. Her grand-father, Schuyler Murdock, while serving a term in the Vermont legislature as the representative of Whitingham, Vermont, was overcome in a fit of despondency and took his own life with a razor. My wife's great-grandfather, Atherton, while shocking oats on his farm in Deerfield, Mass., was shot and killed by an Indian. He had his gun in the field, leaning against a shock of oats, but the Indian got between him and his gun, and shot him. According to best information I can get, this was done about the year 1800.

Elva Lorena Chase, the oldest child and first daughter, was born in Whitingham, Vermont, April 19th, 1876; commenced her schooling in the country school in Whitingham, and finished in the common schools in Pawnee county, Nebraska. Taught school for several years, and married Frank J. Brown, near Pawnee City, December 24th, 1896. Has lived on a farm seven and one-half miles south of Pawnee City ever since her marriage. Has three children and two grand-children.

Mrs. Anna Murdock Chase

Leon Wilson Chase, the second child, was born in Whitingham, Vermont, August 27, 1877, and commenced school in the same town. He graduated from the Pawnee City High School in 1899, and received the degree of Bachelor of Science in Mechanical Engineering from the University of Nebraska in 1904, and in 1914 received a Mechanical Engineering degree from the same University and at the same time an Agricultural Engineering degree from the Iowa State College, this being the first Agricultural Engineering degree ever given at any place. For fourteen years he was at the head of the Department of Agricultural Engineering of the University of Nebraska. With J. B. Davidson, he is author of a text book on Farm Machinery and Farm Motors, the first book of this kind ever written. The following are some of the honorary and professional organizations to which he has been elected: Nebraska Engineers, Sigma Tau, Alpha Zeta, Alpha Gamma Rho, Farm House, Charter Member of American Society of Mechanical Engineers, Society for Promotion of Engineering Education, American Society of Automotive Engineers, and Sons of the American Revolution. He offered his services in the World War and was commissioned Major of Ordnance in 1918, honorably discharged in 1919 and re-commissioned the same year as Major of Ordnance Officers' Reserve Corps, and promoted to Lieutenant Colonel of Ordnance, Officers' Reserve Corps in 1924. Married Susan Mills Barnhart of Logan, Ia., in that town on August 16th, 1905; has three children living. He is now President and General Manager of the Chase Plow Company, of Lincoln, Nebraska, with his residence at 1245 North 33d Street.

Dwight Leslie Chase, the third child, was born in Whitingham, Vermont, November 18th, 1878. Attended the common country schools in Pawnee county, Nebraska, and has specialized in farming. He married Josephine Slack, of South Fork precinct, March 12th, 1912, in Pawnee City, Nebraska, and went right to farming; now lives on a farm in South Fork precinct, Pawnee county.

Carl Fred Chase, the fourth child, was born in the village of Jacksonville, town of Whitingham, county of Windham, Vermont, February 4th, 1881. Graduated from the Pawnee City High School, and from the University of Nebraska, class of 1910. For several years was instructor in the Kansas Agricultural College; resigned and took up the same work in the North Dakota Agricultural College; resigned from here and entered into the real activities of agriculture, and is now managing an irrigated ranch near Nampa, Idaho. Married Estella M. Stephens, of Crete, Nebraska, in Crete April 27th, 1912. Living now in Nampa, Idaho. Has four children.

Floyd Josiah Chase, the fifth child, was born in a dug-out on a farm near Violet, Pawnee county, Nebraska, October 28th, 1886. Attended the Pawnee county schools, graduated from the Pawnee High School and from the University of Nebraska, class of 1914. Was assistant in dairying and farm management at the University of Nebraska for a while, later going to Montana, where he was an extension specialist in farm management. Was taken in the first draft from Montana in the World War, was commissioned sergeant in the 346th Machine Gun Batalion, went across the sea and was in three major offensives of the American Army. When the armistice was signed he was serving with the Belgian Army. After that until the time of returning home he taught farm management at the University Farm at Allery, France. After his return home he was a county agent in Montana for a time, then went to the Agricultural College, Ft. Collins, Colo., where he was a farm management specializer. He is not married. Is now working with the Biological Survey in New Mexico, with headquarters at Albuquerque, care Y. M. C. A.

Charles Murdock Chase, the sixth child, was born in a dug-out near Violet, Pawnee county, Nebraska, September 2d, 1888. Attended the Pawnee county public schools, and one year in the State Agricultural School at Lincoln, Nebraska. Specializes in dairy farming. Married Elsie L. Martin, of Pawnee City, in Pawnee City September 26th, 1911, and is now living on a farm three miles north-east of Pawnee City, Nebraska. Has three children.

Linus Chase, the seventh child, was born in a dug-out near Violet, Nebraska, February 18th, 1890. Attended the Pawnee county schools; graduated from the Pawnee City High School and from the University of Nebraska, class of 1916, and has specialized in dairy farming ever since. Is a member of the Lincoln Chapter, Nebraska Society of the Sons of the American Revolution. Received from the Governor of Nebraska a commission as retired officer of the University Cadets, with the rank of Second Lieutenant in the Reserve Militia. Married Anna Liebers, of Lincoln, Nebraska, in Lincoln, December 29th, 1918. Now living on a farm four miles north-east of Pawnee City, Nebraska, on State Highway No. 26. Has two children.

Lela Ione Chase, the eighth child and second daughter, was born in a dug-out near Violet, Nebraska, February 7th, 1892. Attended the Pawnee county public schools, and was graduated from the Pawnee City High School with the class of 1912. Taught school in the Pawnee county country schools for several years. Married Fred L. Bachenberg, of Lewiston, Nebraska, on the Chase Farm September 23d, 1914, and with her husband has lived on a farm ever since. Now lives three miles west of

Our Present Home in Pawnee City, Nebraska

Pawnee City, Nebraska, on State Highway No. 26. Has three children.

Glenn Harold Chase, the seventh son and ninth child, was born in Norton, Norton county, Kansas, February 27th, 1894. Attended the Pawnee county, Nebraska, public schools. Graduated from the State School of Agriculture in the class of 1916, and specialized in general farming until the Chase Farm was sold, November 13th, 1919; since then he attended the Lincoln Business College for one year, and has been traveling salesman for different implement companies. Is now (October, 1924) head salesman for the Chase Plow Co., of Lincoln, Nebr. Married Sadie Ann Jones, of Pawnee City, Nebraska, in Pawnee City September 14th, 1922. Has one child. Lives in Lincoln, Nebraska, at 534 North 35th street.

We celebrated our golden wedding day, December 24th, 1924, with all of our children, in-laws and grand-children present except Carl's wife and his four children.

A very timely poem, entitled "Fifty Years Together," composed by Mrs. Linnie Faulkner, of Pawnee City, for the celebration, follows in full:

FIFTY YEARS TOGETHER

Way back in New England,
  Eighteen seventy four,
'Twas the eve before Christmas;
  Great logs with a roar
Sent sparks up the chimney
  And lighted the floor.
Before the huge fireplace
  Stood a young farmer lad,
Beside him a lassie—
  Both faces were glad.
Fred Chase and fair Anna,
  His newly made bride,
His eyes dwelt upon her,
  Full of love's tender pride.

With hopes mounting high,
 Hearts light as a feather,
They had started that night
 Life's journey together—
Pledged to stand by each other,
 Come weal or come woe,
Mid-summer's warm sunshine
 Or winter's cold snow.
Life's journey! Ah whither,
 How far would it lead?
But no seer was at hand
 Their future to read;
No magical carpet
 Before them unrolled
To bear them both swiftly
 To riches untold.
But clear and insistent,
 Came a call from the West;
"Come West, my brave children,
 Where you will be blest
With rewards rich and ample,
 If you but do your best.
'Tis youth and 'tis courage,
 Ideals such as yours,
We need in the building
 A state that endures."
So they came to Nebraska,
 Worked early and late
On "Hard Scrabble Ranch,"
 And oft felt that fate
Had dealt them unkindly,
 A full body blow—
'Twas so seldom one reaped,
 And 'twas always sow, sow!
But at last, fate relenting,
 Smiled bright on the pair,
And directed their footsteps
 To a farmstead more fair.
But still there was labor,
 To bring out the charm
Of what we now know
 As the noted Chase Farm.
Resting now from their labors,
 The Chases we see
In a home new and cozy,
 'Mong the best in Pawnee.
And redeemed is the promise,
 Once made by the West,
To reward all her children
 Should they but do their best.
So shoulder to shoulder
 They've stood thus together,
Mid sunshine and rain,
 And all sorts of weather.

Tho the pathway before them
    Was at times dark and steep,
And the going seemed rough
    To their toil-weary feet,
Yet they've traveled together
    These past fifty years,
Sharing life's great adventure,
    It's laughter; it's tears.
Together! What magic
    In that one little word!
To some 'tis the sweetest
    That ever was heard.
It halves all our sorrows,
    It trebles our pleasure,
A boon we all crave,
    A gift without measure—
Just to know that no matter
    How threat'ning the weather
Fate's blows cannot harm
    If we face them together.
Tho the white snow of winter
    May lie on each head,
In each heart spring's warm sunshine
    Still brightly is shed.
For love is eternal,
    No winter it knows;
Sometimes blooming brightest
    When wind coldest blows.
Gathered here are the children
    And grand-children, dear,
While loved friends and neighbors
    Come from both far and near—
Fellow travelers all along
    Life's open highway,
To bid them God speed
    On this anniversary day.
What shall we wish
    For this couple so true?
Wealth, power or position?
    Come now, what say you?
But wait just a moment,
    A thought comes to me,
If you'll ponder it over
    It's beauties you'll see.
Let us wish that each mile-stone,
    Representing this day,
May find them together
    To the end of the way.

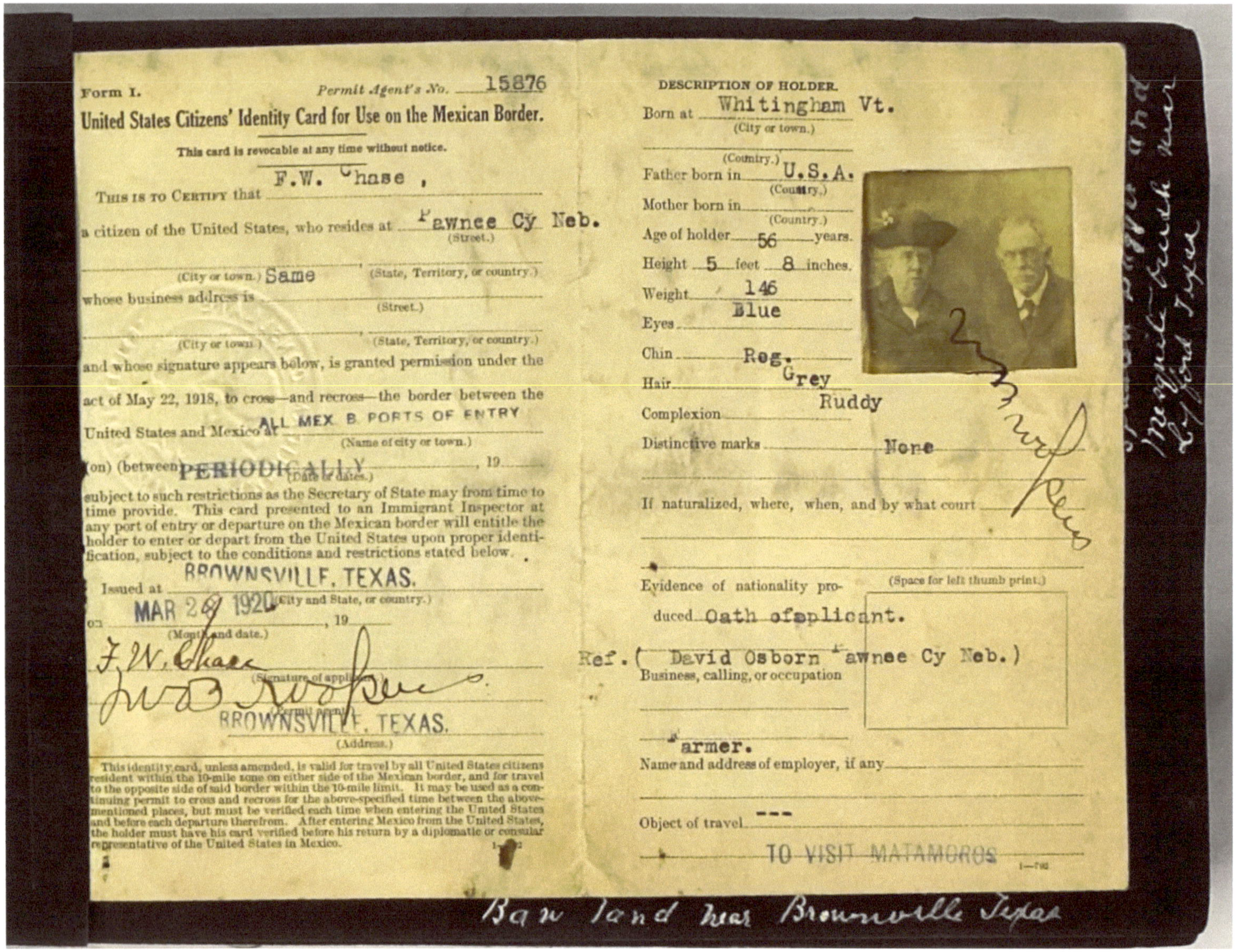

Form I. *Permit Agent's No.* 15876

United States Citizens' Identity Card for Use on the Mexican Border.

This card is revocable at any time without notice.

This is to Certify that F.W. Chase ,

a citizen of the United States, who resides at Pawnee Cy Neb. (Street.)

(City or town.) Same (State, Territory, or country.)

whose business address is (Street.)

(City or town.) (State, Territory, or country.)

and whose signature appears below, is granted permission under the act of May 22, 1918, to cross—and recross—the border between the United States and Mexico at ALL MEX B PORTS OF ENTRY (Name of city or town.)

(on) (between) PERIODICALLY (Date or dates.) , 19

subject to such restrictions as the Secretary of State may from time to time provide. This card presented to an Immigrant Inspector at any port of entry or departure on the Mexican border will entitle the holder to enter or depart from the United States upon proper identification, subject to the conditions and restrictions stated below.

Issued at BROWNSVILLE, TEXAS. (City and State, or country.)

on MAR 29 1920 , 19 (Month and date.)

F. W. Chase (Signature of applicant.)

W. B. Hopkins (Permit agent.)

BROWNSVILLE, TEXAS. (Address.)

This identity card, unless amended, is valid for travel by all United States citizens resident within the 10-mile zone on either side of the Mexican border, and for travel to the opposite side of said border within the 10-mile limit. It may be used as a continuing permit to cross and recross for the above-specified time between the above-mentioned places, but must be verified each time when entering the United States and before each departure therefrom. After entering Mexico from the United States, the holder must have his card verified before his return by a diplomatic or consular representative of the United States in Mexico.

DESCRIPTION OF HOLDER.

Born at Whitingham Vt. (City or town.)

(Country.)

Father born in U.S.A. (Country.)

Mother born in (Country.)

Age of holder 56 years.

Height 5 feet 8 inches.

Weight 146

Eyes Blue

Chin Reg.

Hair Grey

Complexion Ruddy

Distinctive marks None

If naturalized, where, when, and by what court

Evidence of nationality produced Oath of aplicant.

(Space for left thumb print.)

Ref. (David Osborn Pawnee Cy Neb.)

Business, calling, or occupation Farmer.

Name and address of employer, if any

Object of travel ---

TO VISIT MATAMOROS

Ban land near Brownsville Texas

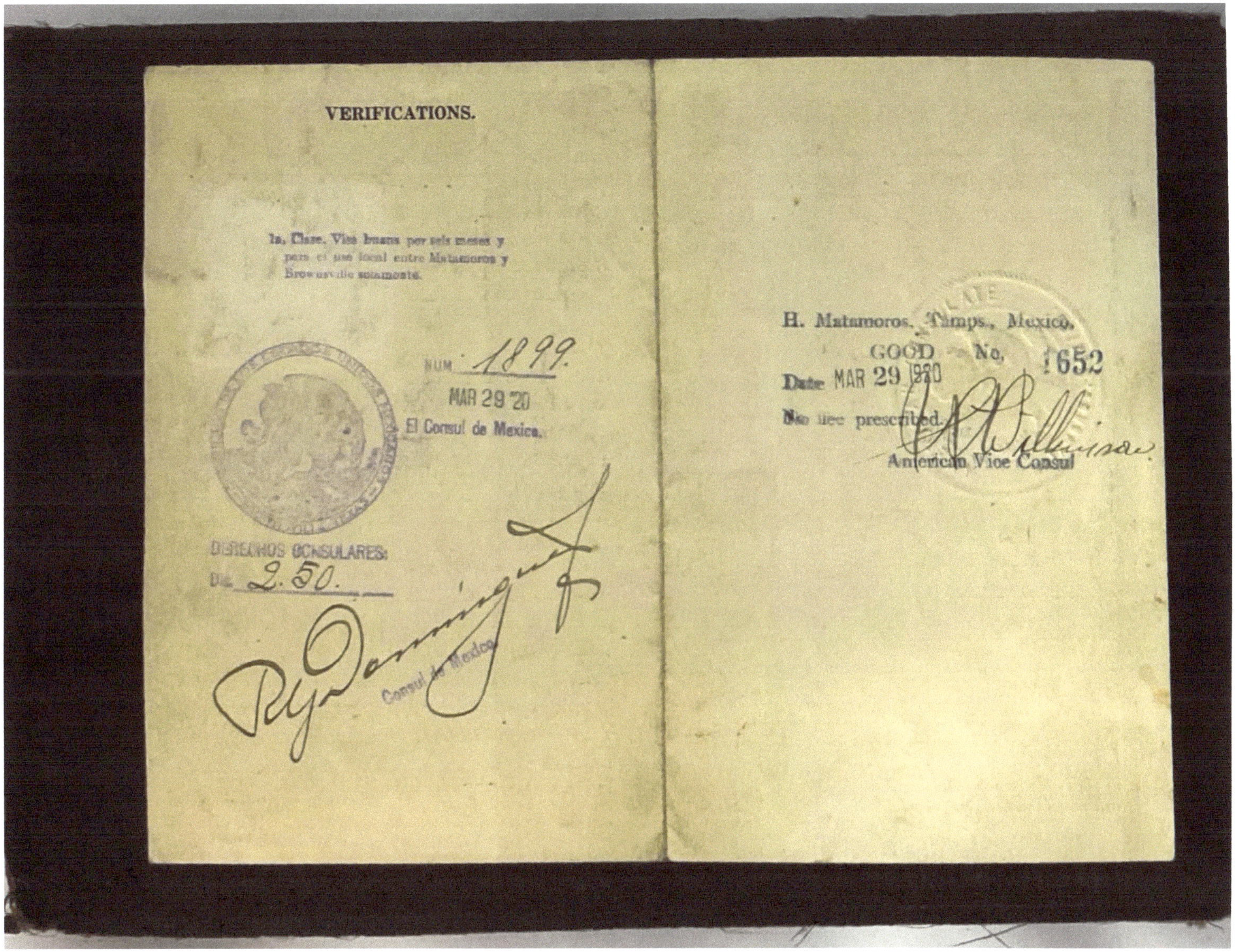
VERIFICATIONS.
1a. Clase. Visa buena por seis meses y
para el uso local entre Matamoros y
Brownsville solamente.
NUM. 1899.
MAR 29 '20
El Consul de Mexico.
DERECHOS CONSULARES:
Dls. 2.50.
Consul de Mexico
H. Matamoros, Tamps., Mexico.
GOOD No. 1652
Date MAR 29 1920
No fee prescribed.
American Vice Consul

Spanish Dagger near Combes Texas

Spanish Dagger and Mesquite brush near Lyford Texas

Raw land near Brownville Texas

Cactus and brush on uncleared land near Brownville Texas

Residence and Palms in Sanbenito Texas

A Mexicans outfit while Clearing near Harlingen Texas

Mexicans Clearing raw land near Harlingen Texas

Mexican homes made of native material. The water barrel has a pin in each end The mexican then takes a rope about 20 feet long ties a ring in each end puts one ring over one pin in the end of the barrel and the other ring over the other pin makes a loop in the center of the rope and hangs it to the saddle and away he goes after water rolling the barrel on the ground

Mexican home near Mercedes Texas made of native material

Mausoleums in Mata-
moros Cemetary old Mexico

Mausoleums in Matamoros cemetary old Mexico

Beggars in Matamoros
old Mexico
note the cactus on the
roof

Matamoros
Cathedral construction
was completed in
the year 1718

Water peddlers filling their barrels in Matamoros old mexico

a water peddler in Matamoros old Mexico The city is one of the oldest on the american continent and has 40.000 inhabitants

Street car system in Matamoros old Mexico

The arena in Matamoros where the bull fights are held seating capacity 10,000 lowest price seat to attend a bull fight is $2.50

Residence street in Matamoros, all windows and doors have steel bars over them

an ox dray in Matamoros Mexico

Cactus trees in Matamoros

a Mexican's broom corn thresher

Sixty acres of onions in Raimondville Texas Palm in front orange trees in field

Mexicans topping onions in Raimondville Texas

Onions crated ready for shipment at Ramondville Texas

Seventy six loads of cabbages waiting their turn to unload at one place in Harlingen Texas Cotton gin in the rear

Unloading cabbages in to the cars at Donner Texas

Unloading cabbages in to the cars at Donner Texas
The loads weigh from one & a half to two tons

Mexican Corrall and ponies

Main canal of the Mercedez irrigation project three miles from the 2nd. lift - or about 13 miles from the 1st. lift Water 80 ft above the Rio Grande River

Pipes at the Donner pumping plant. Donner Texas
By the use of pump and three pipes water is pumped from the Rio Grande river into irrigation & canal 30 ft above water in river
Diam of pipes 48 inches

Rio Grande River
near the Donner
pumping irrigation
plant.
Mr & Mrs Walters
of Casselton N.D.
changing the film

Rio Grande river near
Donner pumping plant
Mr Walters on the bank
The Mexican cabin and
brush are in old Mexico

Oriole Window in the
San Jose Mission
San Antonio Texas

Door way at San Jose
Mission San Antonio Texas
Ecrected in 1715 or about that

Japanese sunken garden
in old rock quarry at
Breckenridge Park San
Antonio Texas
See the cranes

Banana tree and bud
1st leaf of bud shed and
little bananas forming

Orange trees and fruit Brownville Texas

Oranges trees and fruit Brownville Texas

Mrs Katie E. Houghton
Lyford Texas
Major Everett Collins
Fort Sam Houston Texas
Mrs Anna M. Chase
Pawnee City Nebr
Taken in Harlingen
Texas

New Bank Block
San Benito Texas
architectural design
Spanish

E. L. Houghton's residence Lyford Texas. Front and rear view. April 1920

E. L. Houghton filling his water barrels from artesian Well Lyford Texas, april 1920

Mr and Mrs E. L. Houghton and grandson Ledoit Yundt Lyford Texas april 1920

Lidoit and Letha Yundt at left and Francis and Helen Houghton at right, Grand-children of Mr & Mrs. E. L. Houghton Lyford Texas about April 1st 1920

Mrs Elsie Yundt and Mrs Ruby Houghton Lyford Texas about Apr 1st 1920

Alta and Clara Houghton
Lyford Texas
about April 1st 1920

Elva Yundt and

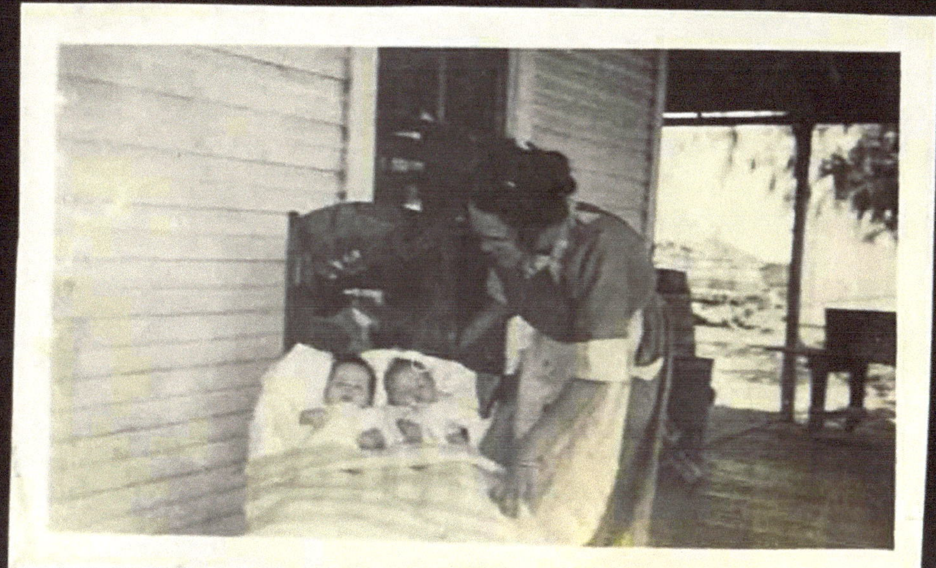

150 bushels of sweet potatoes damaged by weavel near sorting shed in Harlingen Texas

Lyford Texas

Tamworth hogs on the Dryo and Ray ranch near Lyford Texas Wt 400 each Ray Houghton standing

Bridge over the arroyo between Harlingen and San Benito Texas 40 ft above low water

The Cabildo
New Orleans

This building was erected in 1795 was the Spanish Court house in Colonial days and some of the implements of torture used at that time are still preserved in the building. a noticable one was a set of heavy iron bound stocks The court in the rear of the building was the scene of many executions. The room in the upper left hand corner is where the transfer of Lousiana from Spain to France and from France to the U.S. was made, Lafayette had his quarters in the Cabildo in 1825

Took a three hour trip in the Sidney up and down the Mississippi river

Statue Gen. Jackson
New Orleans

Statue Robt E. Lee
New Orleans

Tomb Stone in Cemetary New Orleans Cost $85,000

Mausoleums in New Orleans Cemetary

Residence of Major Everett Collins Fort Sam Houston near San Antonio Texas

a ship on the Mississippi river Exposure made from an other moving ship

The largest sugar refinery in the World. Located below New Orleans. Exposure made from a moving ship

Fords Theater where President Lincoln was assinated by J. Wilkes Booth on the night of april 14th 1865

New Orleans special running daily between New Orleans and New york city. Pullman service entire.

X was our car

U. S. Capitol Building
Washington D.C.

The White House
Exposure made
through the iron
fence that surrounds
the grounds

Washington Monument
550 ft high
Gen Sherman's Monument
Washington D C
Mrs Chase Mattie Stewart
and Veryl Hall standing
Garfield Statue
Washington D.C.

Rear view of Mt Vernon

Front view of Mt Vernon

D.A.R. Building
Washington D.C.

Steps in rear of
Capitol Building
Washington D.C.

Fort Washington on the Potomac river between Washington and Mt Vernon

U.S. Treasury Building Washington D.C.

White House
Exposure made from top of Washington Monument

×Capitol Building and ×Nat[l] Museum
Exposure made from top of Washington Monument

Out-buildings at Mt Vernon

Cadets training near Washington Monument

Steam shovel in operation
in Greenfield Mass
Exposure made when the
shovel was in operation

Shovel turned and
is now emptying

Church in
Williamstown Mass

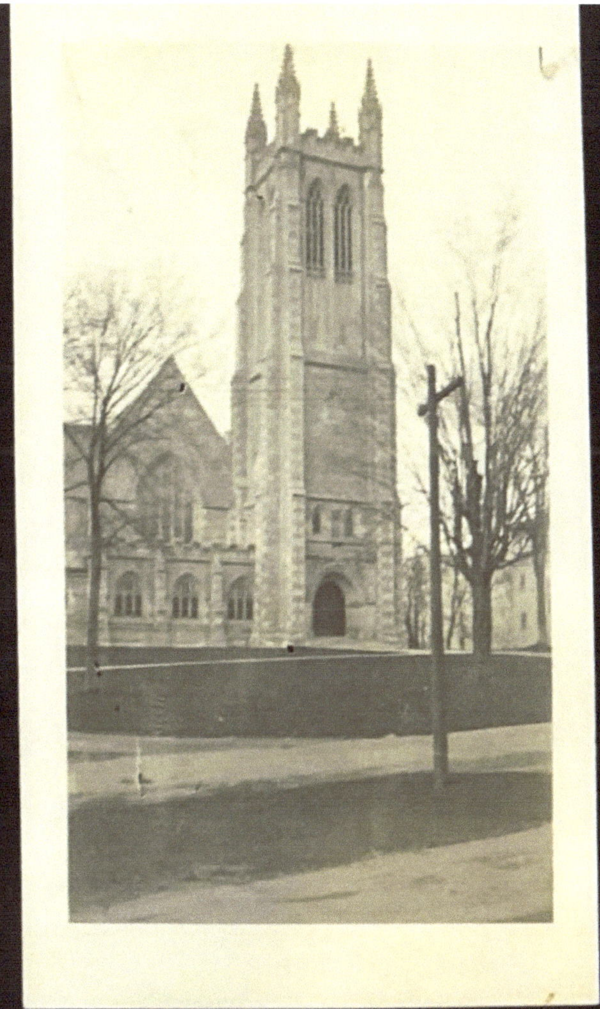

Memorial Hall
Williamstown Mass

Chapel Building
Williamstown Mass
made of cobble stones

College Buildings
Williamstown Mass

Main St Williamstown ~~Vt~~ Mass

Auction day on the Goodnow Farm Whitingham Vt May 12 1920

+ Snow banks in the Whitingham Vermont hills May 12 1920
Hay stack Mountain in the distance
Whitingham or Sadawga lake center on the left
Stone wall fence on the right

+ Snow bank on the farm that Anna M. Chase was born Whitingham Vt.
Exposure made May 12th 1920

Mr and Mrs.
H. L. Chase
North Adams Mass

Mrs Mable Wheeler
and three children
Marjorie. Irene.
Pauline. Home in
Shelburne Mass
Exposure made May 12
1920 at Goodnow auction
Whitingham Vt

On The Mohawk Trail
a very steep winding mac-
cadam surfaced. road over
the Hoosac mountain in
Mass. 16 miles long from the
foot of the mountain on
one side to the foot on
the other side raising in

height in one place to
over 2100 feet and cost
the State of Massachusetts
$340,000 The inclines are
so long and steep it is
not considered profitable
for commercial Trafic

Catching a drink on the Mohawk Trail.
There are several of these springs and watering places on the trail

Concrete bridge and babbling brook beside the Mohawk trail

Vermont hills.

Exposure made 1600 feet above the river in the Deerfield Valley. below.

Exposure made from the Mohawk trail. mile stone at right

Stamford Vt. valley from the hairpin bend on the Mohawk Trail

Bridge over the Deerfield river on the "Mohawk Trail"

Cheshire Lake & Farnams lime works Cheshire Ma

Lime Works near South Adams Mass
x The long streak of White at the foot of the Mountain is lime rock quarries. Exposure made one mile away

This boulder slid from the side of the mountain and entirely obstructed the road from travel near Readsboro Vt. several work men were employed three days blasting off one end ~~so~~ sufficient so that travel could be resumed
Exposure made Apr 25 1920

Mrs Albert Wiley and her home in Buckland Mass

Buildings on "River-side" farm near Jacksonville Vt Mr and Mrs Forest Barker's home

Charles Barker and his buckboard, Forest Barker standing with foot on wheel May 1920

Forest Barker trimming rose bushes and his wife looking on. "River side Farm" Jacksonville Vt. May 1920

Forest Barker of "River-side Farm" just returning from taking the cows to pasture The long whip takes the place of the Scotch Collie

Mrs Harry Murdock and two children Catherin Grace and Ramond Harry. and Mrs Forest Barker Jacksonville Vt

Old "Buff" the domestic on "River-side Farm"

Major 27 years old
For 24 years Major has been producing power to carry on Riverside Farm and has always been stabled since 3 years of age has not ~~never~~ been in pasture since

Rear view of River Side Farm Jacksonville Vt

View in Albert Faulkner's cow pasture near Jacksonville Vt.

Corn crib on River-Side Farm near Jacksonville Vt

Cultivating corn on W. H. Hancock's farm May 2d 1920 near Jacksonville Vt

In Whitingham Cemetary
Decoration day May 30th
1920
Tomb stones of the
father and Mother of
F. W. Chase

In Whitingham cemetary
Decoration day May 30th
1920
Monument to the Memory
of the Father and Mother
of Mrs Anna M. Chase

Tomb stone of Samuel Parker and third wife in Whitingham Cemetary May 30th 1920
Samuel Parker was a soldier in the Revolution and great Grand father to F. W. Chase

The Davidson Cemetary Whitingham Vt May 30th 1920 Moses Chase Grandfather of F. W. Chase is burried here. also Schuler Murdock and Calib Murdock. Grandfather and Great grandfather of Mrs Anna Murdock Chase

Four Murdock Sisters
Effie I. Barker
Jacksonville Vt
Carrie S Plumb
North Dana Mass
Jane F. Dix
Greenfield Mass
Anna M. Chase
Pawnee City Nebr.
Exposure made in Jacksonville
Vermont May 31st 3 oclock P.M. 1920

Annie Hall Chase died at the age of 94 having lived in this house 75 consecutive years lacking 6 days.

The house itself has long since passed the century

Mark

Amy Chase Tinney standing grand-aughter of Hiram Chase now occupies the old homestead

June 9th 1920 By F.W. Chase

April 23d 1835 Hiram Chase (born Oct 9 1812) married Annie Hall and on that date went to houskeeping in this house located in Marlboro Vt. For four years this worthy couple was without matches or timpiece. Fires were started with powder and flax lint, or tow, or coals were borrowed from a neighbor. When their first child was born Annie requested that some means should be provided that a light could be started at once. So a box of matches containing 12 was procured costing 10 cents, they lasted two years and 2 out of the twelve failed to ignite

This house was built in 1834 by Moses Chase. J. B. Chase was born on this farm apr. 10 1826 and died in this house July 14th 1888. F. W. Chase was born in this house June 16th 1853.
June 12th 1920

This is the birthplace of Elva Lorena Lion Wilson and Dwight Leslie Chase
June 12th 1920

C. F. Chase of Nampa Idaho was born in this house

It was on this spot that Elva L. Chase Brown and Leon W. Chase commenced to attend school Their father F. W. Chase and grandfather Josiah Briggs Chase also at one time attended school on this ground.
June 12th 1920

Rena Faulkner Cutting and her collie dog the indespensible on the Cutting Farm an old shaving horse for making shingles is her seat June 9th 1920

Charles Cutting and his maple sap gathering out.
His snowshoes stand at the right
June 9th 1920

These buildings are on the Cutting Farm near Jacksonville Vt. June 9th 1920

Planting potatoes on the Cutting Farm. Charles Cutting is droping the commercial fertiliger and his son Donald is covering the potatoes and fertiliger with the hoe June 10 1920

A tractor in service on what is known as the Hiram Chase farm in Marlboro Vt now owned by his granddaughter Amy Chase Tinney note the team and man hauling the loose stone away plowed out by the tractor June 9th 1920

The tractor going over the rock at the right is the same tractor as the one above and is in the same field
June 1920

Droping commercial fertiliger and planting corn on a Whitingham Vt farm
This is on a good Whitingham farm

View of Floating land on Sadawga lake Whitingham Vt

C. H. Waste of Jacksonville Vt at the tomb stone of his great great grand mother Lucy Harrington in the Whitingham Vt cemetary

These apple trees were grafted in the spring of 1875 by F. W. Chase. Exposure made June 12th 1920 Forty five years after the work of grafting was done.

Scene among the Whitingham
Vt hills June 8 1920
Jacksonville Vt school house

Jacksonville Vt pond
from A.J. Faulkners
hill June 8 1920

Forerivor ship building yard Quincey Mass June 23rd 1920

Atlantic ocean near Plymouth Mass June 23 1920

Five tobacco curing barns each 100 ft long.
The tobacco is cut in the field and strung on to four feet lath and put in to these barns by suspending on girts. Tobacco tent at the right.

35 acre tobacco field under tent.
The cost of this method of covering tobacco is 500.00 a season per acre
It has the advantage of protecting the plant from hail and makes a much more valuable product

a tented tobacco field.
Tobacco just set
Exposure made under cloth.

The same field as above
looking in a different
direction
35 acres all under cloth
Exposure made under the
cloth

W. S. Pinney's residence
Suffield Conn
We lodged here with Marcus
the night of June 19th 1920
This is on the George
Washington trail from
New York to Boston
Mr. Pinney in the picture is the
owner off the tobacco fields
in this series

on the Washington Trail
near the above residence
a small Inn was located
under this elm and
Washington was known to
have made this Inn one of
his stopping places while
enroute to Boston
Exposure made June 20th 1920

tent not up

In a 35 acre tobacco field now in potatoes, considered a good rotation in tobacco culture
June 20 1920

A tented 15 acre tobacco field from the Pinney barn

Exposure made from Pinney barn

A landscape view of the Conn valley from a Knoll near Suffield Conn June 20th 1920

Remnant of old New-gate prison Near Suffield Conn June 20 1920

New gate prison near Suffield

Remnant of Newgate prison
near Suffield Conn
June 20 1920

Remnant of Newgate prison
near Suffield Conn
June 20 1920

Canal lock or gate at Windsor Locks Conn June 20th 1920

Same canal as above and two gates June 20 1920

Swan boats in the Boston public garden June 22 1920

George Washington Statue in Boston public garden June 22 1920

Sagamore bridge across
Cape Cod canal at Sagamore
Cape Cod canal is a canall
across the cape to let ocean
steamers through from N.Y.
to Boston instead of going
around the point of cape
June 25 1920

Sagamore bridge open
to let a ship through
see the ship in the
canal
June 25 1920

Holyoke Bridge 1/4 mile long. across the Conn river at Holyoke Mass June 19/20

Holyoke dam made from the above bridge June 19 1920

Marble tablet on
Plymouth Monument
June 23 1920

Plymouth Monument
Plymouth Mass
June 23 1920

Marcus and Plymouth rock
June 23 1920

Mrs Florence Lawrence and children and F.W. and Wife at Plymouth rock
June 23 1920

Excursion boats landing at Plymouth from Boston landing at Plymouth rock
June 23 1920

Church in Quincy Mass John Quincy and John Quincy Adams two Presidents of the United States are burried under the porch of this church

Exposure made June 23 1920

Cablegram receiving station
East Orleans Mass on Cape
Cod
All out going and in coming
cablegrams from France to
the United States is sent
out and received in this
house
June 24 1920

Ivy Lodge Inn East Orleans
Mass on Cape Cod
We lodged at this Inn
June 23 1920

at Orleans

We had our supper here June 23 and breakfast June 24 1920..

A cranberry bog on Cape Cod. This land is Valued at 500.00 an acre usually produces 75 barrels of cranberries a season

This stone marks the first landing place of the Pilgrims Nov 11 1620 It was here that the Pilgrims first touched foot on American soil Provincetown Mass extreme point of Cape Cod June 24 1920

This Advertisement speaks for it self June 24 1920

These little boats are fishermens boats in Cape Cod bay off the beach at Provincetown, Cape Cod Mass
Fish guano a fertilizer made entirely of fish is quite an industry at this place
June 24 1920

This ad speaks for it self
June 24 1920
Truro is eight miles from Provincetown Mass

Wind Mill at Chatham Mass on Cape Cod

an old old power wind
Mill on Cape Cod
June 24 1920

Washing her hands in a
~~the~~ neck of the atlantic
ocean
June 24 1920

near Chatham Mass

Here is where we lodged
the night of June 24th 1920
The part marked × is the original
part that was built in early
1700 and has been used for
a hotel ever since. Daniel
Webster often put up here
hence its name.
In the town of Sandwich Mass

The Mayflower Food shop
on Cape Cod
Took a lunch here the
P.M. of June 24 1920

Toy land on Cape Cod.
I was from here that Marcus
sent Marcus Donald a little
Windmill
June 24 1920

French freighter in Charlestown harbor

French freighter in
Charlestown Mass harbor
Bunkerhill monument
in the distance
June 26 1920

U.S. war vessel in
Charlestown Harbor Mass
June 26 1920

U.S. war vessel

The old home of Paul Revere in Boston Mass. Preserved now as a memorial
June 26 1920

The signal lanterns of Paul Revere displayed in the steeple of this church april 18th 1775 warned the country of the march of the British troops to Lexington and Concord
June 26 1920

The exposure for this picture was made from the spot where the first blood of the american revolution was spilled.

The first colonial State house of Mass
From a window in the rear of this building George Washington reviewed his troops.
June 26 1920

Baxter Newells home
Wollaston Mass
We staid here the night of June 27 1920

Revere Beach June 27th
1920

Revere Beach
June 27 1920

The Boston Floating Hospital in Charlestown Harbor Mass
June 26 1920

Spot Pond Stoneham Mass
June 26 1920

Beacon St Boston Mass
a residence street
June 26 1920

In order to take a picture of these mounted cannons in Wattertown arsinal Mass I had to get a permit from the presiding U.S. General
June 26 1920

a large Machine in Wattertown Arsenal Marcus Chase Standing.

Main St in Northfield
Mass
July 4 1920

Kenarden Hall Northfield
Mass
July 4 1920

Mrs Russel Sage chapel
and Helen Gould Hall
Northfield Mass
July 4 1920

The Moody Tabernacle
Northfield Mass
July 4 1920

Northhampton is the home of Calvin Coolidge who was elected Governor of Mass by the largest vote ever recorded for that office and is now the Republican nominee for Vice President July 9 1920

Old conn trail in Weston Mass

This is us eating our dinner July 4 1920 on a rock by the road side a few miles north of Rutland Vt Ethel Plumb snapped the Kodak

Morrill Hall Agricultural College Burlington Vt July 5 1920

Montpelier
United States Tires
Are Good Tires

Bellows Falls
United States Tires
Are Good Tires

Warren E. Waste's
Residence
Greenfield
Mass

Warren E Waste
in his garden

Clayton Plumbs
Rubber Store
Northhampton
Mass

Power Co. Station
Vernon Vermont
Conn. River

Crossley Hall
Mount Herman
School for boys
only. four miles
from Northfield
This Hall is Most for rooms
only

Chappel at Mt
Herman School

A part of the 200 head of Holstein cattle at the Mt Hermon school Mt Hermon Mass

Cutting rye with a cradle on Greenfield Meadows Mass

Stockbridge Hall
Amherst Agricultural
College Mass.
Erected MCXXXIV

Stone wall fences
and meadows near
Hardwick Mass
Hardwick is noted for
its famous Guernsey
and Jersey cattle
Several herds number
200 head

Carl Wheeler son-in-law of Adin Plumb standing

Cliff on Mt tom near Holyoke Mass

A continuation of the cliff on the left with Mt tom observatory on the top

Observatory on Mt Tom
near Holyoke Mass
1218 ft above sea level

onion and tobacco growing district

Exposure made from Mt Tom
View of farms in the valley
at the foot of Mt Tom

Holyoke Mass and Conn river four miles from Mt Tom
Exposure made from Mt Tom

Northhampton Mass ten miles from Mt Tom
Exposure made from Summit of Mt Tom

Park and boulder at foot of Mt Tom

This Tavern is located in West Brookfield Mass and is noted for being one of the stopping places of Geo. Washington while on his trips from Washington to Boston Mass

Vermont Capitol
at Montpelier Vermont
July 5 1920

A railroad wreck
that we passed just
above Bellows Falls
Vt while on our trip
July 4 1920
from North Dana Mass
to Burlington Vt

an abandoned meadow
on the Riverside Farm
near Jacksonville Vt
now used for pasture-
apple trees showing

Hill Lot
River Side Farm

I guess I can mow some myself

"Old Major" Is Dead.

After 29 years of faithful service in the family of F. C. Barker of Jacksonville, "Old Major" one of the best known carriage horses in this section, is dead. "Old Major" was 32 years old and during the 29 years he had been in Mr. Barker's possession had not known a single day of illness nor missed a meal. Up to this winter he had been able to outdistance many horses of younger age, but during the past few weeks his age had begun to weigh heavily upon him until, finally, he surrendered the fight. Mr. Barker and his family as well as all residents of Jacksonville, who were fond of "Old Major," will miss him.

Earl Purington and his oxen. Buckland Mass

July 16 1920

Where this monument stands Capt. Thomas Lathrop and 84 men including 18 teamsters from Deerfield Mass conveying stores from that town to Hadley were ambuscaded by 700 Indians and the Captain and 76 men were slain Sept 18th 1775 The water in the brook ran red with blood

Fountain In Boston Common

Aunt Carrie, Aunt Jane, Mother,
Aunt Effie,

Cousin Bertha Faulkner
Dary and her baby

Uncle Clifton
AND
His Three Children.
Malah, Harry, Amy.

Two Reo Cars Standing In Front Of First National Bank GARDNER Mass Front car owned by Marcus Chase of Wellsley Mass Rear by F. W. Chase of Pawnee City Nebr.

Wendall Bridge

Wendall Mass.

East End of the Wayside Inn

The Wayside Inn
Was The Home of Longfell-
ow.
It was here that his
celebrated poem The
Wayside Inn was written

A dirt dam in Somerset
Vt. This dam is over one
half mile long. 120 ft
high. 100 ft broad on top
makes a resivoir fourteen
miles long and was constru-
cted at a cost of nearly
three millions

Sadawga Lake
Whitingham Vt
Glenn Chase and His
mother in the little boat
in the distance

Mother looking for Witch-hazel blossoms near Pitsfield Mass.

Jacksonville Vt. Band marching to the cemetary decoration day May 1920

In 1686 a house on this Lot was presented to Rev. Johon Williams the first minister of Deerfield Mass. In March 1704 the family was captured & the house burned by Indians Those of the family that were not murdered was marched to Canada In 1707 this house was erected & presented to Rev. Williams who had been released from captivity and had returned to Deerfield
Exposure made Sept 1920

Brown Bros. Sexaphone Sextette Band of Springfield Mass. Playing in Wilmington Vt Old home Week Aug 1920

Band and Procession marching in Wilmington Street Parade day old home coming week Aug 16th 1920

A part of the parade in the Sts of Wilmington Vt old home coming week Aug 1920

Gov Percival Clement of Vt and Oscar Butterfield of Wilmington Vt

Gov. Percival Clement of Vt delivering an address in Wilmington old home week 1920

Hoosac Valley one mile from North Adams Mass.

A section of Hoosac Valley near North Adams Mass. Note the rocks. ×
Greylock Mtn. The highest point in Mass at the left

Left to right
O. J. Hale age 78 Guilford Vt
Orson Thayer " 92 "
Mrs. Bishop " 90 "
Mrs Hale " 80 "

Mrs Huling the
center of this picture
is 94
Mr Thayer Mrs Bishop
Mrs Huling and Mrs
Hale are brother and
sisters combined age 356
years

In The Granite
Quarries Fitzwilliams
N.H.

In the granite quarries Fitzwilliam N.H.

Rear View of Mrs Marion Williams summer cottage at Dublin N.H.

Cost $80,000

Front View of above Cottage

This Auto. is owned by Mrs Marion Williams a 90 horse power Mercedies made in Germany. used by Kaiser "Bill" in 1912 to view his army. cost Mrs Williams $14500 in Germany seats only four people the driver the porter the maid and Mrs Williams

Senator McVeigh's Orchard near Dublin N.H.

Masons Working on
Proff.r Pumpellie's Cottage
Dublin, N.H.

Monadnock Lake
From Prof.r Pumpellie's
Cottage

Camp Oahe

Near Keene N.H.

Forty girls camp here during the summer months in charge of Mr. & Mrs Dr. Eastman / Dr Eastman is a full blood Indian & a Lawyer His home is in Washington D.C.

Lake in Front of Camp Oahe

Granite Lake N.H.

Dublin N.H.

This Summer Cottage at one time was owned by Sec. of the Interior Hitchcock under Roosevelt now owned by a man in St Louis

DUBLIN
11 MILES FROM HERE
United States Tires
Are Good Tires

EENE
United States Tires
Are Good Tires

Mt. Monadnock
Dublin N.H.
3186 ft above sea level.

Mt Monadnock

Dublin N. H.

Lake Monadnock, has no inlet, 1605 ft. above sea. Exposure made from near the summer home of Abbott H Thayer the noted Painter

Neil Duncan's home
Morrisville N.Y.

New York Educational
building Albany N.Y.
Exposure made in the
rain

New York State
capital building
Albany N.Y.
Exposure made in the
rain

Looking down Niagara river from goat island

Niagara Falls from the bridge below the Falls.

Horse Shoe falls from Goat Island

The Niagara River above the falls looking towards Canada from Goat Island

Rapids above Goat Island Bridge on the America Side

Niagara River above the falls between goat Island and N.Y. Shore

Niagra Falls from the N.Y. Shore

Niagara Falls from the bridge below

Alice, Charles, Mother and Glenn in W. A. Tickner's green house Ashtabula Ohio

Mr & Mrs W. A. Tickner in their green house Ashtabula Ohio

Mr + Mrs W. A. Tickner Ashtabula, Ohio

Our Camp near Niagara Falls

Same as above

A Scene in Indiana

Lake Michigan
from Lincoln Park
Chicago

Geneva Wisconsin

dinner near Orion Ill

Our noon day lunch consisted of a broken watermelon, "hot dogs" and bread near Orion Ill.

View of Ill land from the cupalo of the McConnell barn near Alpha Ill.

Bridge across the Mississippi River at Rock Island Ill.

Some of the University buildings at Iowa City Iowa

TAMA
8½ MILES FROM HERE
United States Tires
Are Good Tires

HISTORY OF THE UNITED STATES
AMES
MILES FROM HERE
HISTORY OF THE UNITED STATES
United States Tires
Are Good Tires

State
Agricultural Engineering
Building at Ames Iowa

State Dairy Building
at Ames Iowa

HISTORY OF THE UNITED STATES
Carroll
3 MILES FROM HERE
Carroll was named for Charles Carroll, a signer of the Declaration of Independence, who, in 1776 was one of the commissioners sent to persuade Canada to join the War of Independence.
HISTORY OF THE UNITED STATES
United States Tires
Are Good Tires

Adjt. General Faulkner's Headquarters. Camp Brown Pawnee Nebr.

Gen. C. A. Faulkner and Capt Clifford Brown camp Brown Pawnee Nebr

ELVA'S HOME

LUCELE

Lela's Home

Corn Shelling Day
at Fred Bachenberg's
Rockford Nebr.

Leon's Home

Leon's Farm

Consolidated Shool Building, Lewiston Nebr.

Corn Husking Machine near Virginia Nebr.

Blue River, Bridge,
Dam, and Flouring Mill
Beatrice Nebr.

Burlington Switch
Yards, Lincoln Nebr.

On the golf links
Lincoln Nebr.
C. A. Faulkner making
the hit Leon W Chase
standing

Golf Field Lincoln
Nebr.

Mr and Mrs. C. A. Faulkner ready to start for their home in Vermont From Lincoln

Leona and Eleanor

Mary Brown

Agricultural Engineering
Building State Farm
Lincoln Nebr

Land scene from car window in Northern Texas on Rock Island train Jan 20th 1920

a country scene from a car window on the Rock Island near Guymore Okla Jan 20 1921

Not far from El Paso Texas the morning of Jan 21st 1921

Hill near Lordsburg New Mexico From the car window Jan 21st 1921

Sage brush in New Mexico from the car window Jan 21 192[illegible]

Sage brush and hills in New Mexico from car window Jan 21st 1921

Going up a steep grade in New Mexico with two engines Exposure made from rear end of train while going Jan 21st 1921

At Colton Calif. on the morning of Jan 22d 1921 Mrs Porter, Louise Crampton and Mrs. Chase

Mountains from the car window in New Mexico Jan 21st 1921

Snow covered Mts in California Jan 24 1921. Exposure made from south of Ontario Calif. in a barley field

San Gabriel Mts Cal.

Mr C.C. Grier's home
in Ontario California
Jan 24th 1921

River Side Calif.
Mr C.C. Grier of
Ontairo standing

A pair of pack mules in Ontario Calif just in from the Mts. in Nevada Jan 24 1920

San Ta Fee depot in San Barnido Calif.

The cross on
Roubidoux hill
Calif

The same as above
only a poor exposure

Riverside Calif.
From the top of
Roubidoux Hill
Jan, 24th 1921

Calif. Agricultural
College near Riverside
from the top of
Roubidoux Hill.
Jan 24th 1921

Valley and mountains from top of Roubidoux Hill California Jan 24 1921

A scene in South Pasadena Calif

The Home of F. N.
Warren Pasadenia
California
F. N. Warren and

A scene in Pasadenia
Calif.

The home of Wrigley
(The chewing gum man)
Pasadenia Calif.

Mrs J. W. Bullards home Long Beach Calif

Mrs A R Stettlar's home Long Beach Calif.

Rear view of Tom Dinnie home Orange Calif.

The home of Tom Dinnie Orange Calif.

Mrs Carrie Dinnis
Mrs. Anna M. Chase
Mrs Ellen M. Hall
Miss Edna Hall

A well laden lemon tree on the Dinnis ranch Orange Calif

Ocean front Long Beach California

Mrs F.W. Chase hunting sea shells Long Beach Cal.

Gathering sea shells
~~near Seal Beach Calif~~
on Belmont Shore Calif
Feb 2d 1921

Gathering Sea Shells
on Belmont Shore
Feb 2d 1921

Trees in a front yard Santa Ana Calif

A vineyard near Lomita Calif.

Fumigating to destroy Orange scale

Ten acre orange grove on Chapman Ave. Orange Cal. owner refused six thousand an acre for it

Mr and Mrs. Leon Houghton
and baby Lomita Calif

Mr and Mrs Leon Houghton

Sea Gulls Rodonda
Beach Calif
Feb 21st 1921

Sea Gulls at Rodonda
Beach Calif
Feb 21st 1921

Playing Rogue Public park Long Beach Calif.

A rogue ground in the Public park Long Beach Calif.

Horse shoe pitching
public park Long Beach

Market day, Long Beach

Feb 19th 1921
The Los Angles oil
Refinery Los Angeles Cal.

Mar 2nd 1921
Oil storage Tanks Puente
Calif.

Mar 2d 1921
Oil wells and pumping derricks in the Puente oil fields Calif

Mar 2d 1921
Oil derricks and wells Puente Calif.

Los Angles Harbor and breakwater wall.
Feb 21 1921

Ship building yards in Los Angeles Harbor
Feb 21 1921

The Pacific Fleet in
Los Angeles Harbor
Mar 8th 1921

Lumber docks in the
distance in Los Angeles
Harbor
Feb 21st 1921

Wilmington Harbor and ship yards near Long Beach Calif

Lincoln monument and Public Library in the Public Park Long Beach Calif.

Looking down the Pike from the end of the pier at the foot of Pine St Long Beach Calif.

A group of bare Skins on Long Beach

4 silos on Los Angeles Co
farm capacity 1400 tons

100 Holestein cows on the Los
Angeles Co. farm for its poor
people. 809 were being cared
for Feb 27th 1921

Feb 27 1921

Cucumbers under cloth in the vegetable gardens of the Japanese. Long Beach Calif. These Japs often raise one thousand dollars worth of vegetables in one season on one acre

Seventy eight seals on the beach. these seals get to weigh four hundred pounds quite often

SEALS

Seals on the beach at Seal Beach Calif. Mr and Mrs D. E. Wherry and Mrs F. W. Chase on the wharf.

Seals on the beach at Seal Beach Cal.

A dead seal on the beach at Seal Beach Cal.

Looking down Pine St from the end of the Pier Long Beach Calif.

Air Plane hangers Daugherty aviation field Long Beach Calif.

A U.S. Gun boat in
Los Angeles Harbor
Taken from the deck
of the Avilon
Mar 15 1921

Light House and
breaker wall Los-
Angeles Harbor Calif.
Mar 15 1921

The Avalon
We took this ship
at the Wilmington pier
Los Angeles harbor and
went 30 miles out into
the Pacific ocean to
visit the Catalina Island
Mar 15th 1921

A scene on the Catalina
Island Mar. 15 1921. From
this pier we took a glass
bottom boat and went
to see the submarine
garden off the coast of
Catalina

Caralina Island from the ship Avalon march 15 1921

A Caralina Island scene off the deck of the Avalon Mar 15th 1921

A U.S. Gun Boat in Los Angeles harbor Calif. Exposure made from the deck of the Avalon Mar 15th 1921

The Pacific at Hermosa beach Calif. Mar 19 1921 3.00 p.m.

The ocean at Hermosa 3/19

Nebr. Picnic at Sycamore Park Los Angeles Cal. Mar 17th 1921

Auto's at the Nebr. picnic Sycamore park Los Angeles Cal Mar. 17th 1921

On Mrs Bullards
porch Long Beach Calif.
Mar 20th 1921

at the entrance of
Topango Canon from
Santa Monica Calif.
Mar 19 1921

On the U.S. Gun boat Wyoming in Los Angeles harbor Mar 21 1921

On the Wyoming Mar 21st 1921

F. W. Chase & John Fournell Mar 22d 1921. a catch off the music near Catalina Island Cal. 45 lbs one eight-hour trip

Sailors practicing rowing in Los Angeles Harbor Calif. Exposure made from a shore boat of the U.S. gun boat Wyoming Mar 21 1921

Resting after above practice

Camels in Barnes circus parade Long Beach Calif Mar 18 1920

Elephants in Barnes circus parade Long Beach Calif. Mar 18th 1920

An English walnut grove near Compton Calif. Ditches have been prepared ready for irrigation

One Thousand hens owned and cared for by W. H. Frost husband of Alpha Miles Frost Compton Calif.

DRYBOROUGH HALL
The Home of Mr.
Dryborough, in Flint
Ridge Calif.
Siera Madre Mts. in
distance

Training Paddock for
Horses on the Flint
Ridge Club Grounds
Flint Ridge Calif.

The Eagle Rock near the town of Eagle Rock Calif It was here we attended the Easter morning sunrise service Easter morn 1921

a Japanese vegtable and fruit market by the highway near Pasadena Calif.

At the foot of the Tehachape range on the ridge route from Pasadena to Bakersfield and the north Calif. April 1st 1921

A scene in the Tehachape mountains from where we eat our dinner on the ridge route between Pasadena and Bakersfield Calif April 1st 1921

Adirondac Wuteke Dairy Maid.
Owned by the Bridgford Holstein Co. Patterson Calif.
This cow weighs a ton and is the fourth largest producing cow in the World

x Sydney Wymores home near Porterville Calif.
Orange groves in the fore ground

On the Ostrich Farm
South Pasadenia Calif.
296 Ostriches were on
the farm

The End Of the Trail
Ira Caldwell Standing

A 75 Horse power Tractor cultivating in the field of H. Jons Paterson Calif. Cultivates 34 feet wide and Works 75 acres a day. In plowing with this tractor 15 ten inch bottoms are used Harry Bull standing at left April 15th 1921

In the hills on the Tunnel Highway Three miles east of Berkley Calif.

Mrs Sawdey and her Rose bush Patterson Calif. A next door neighbor to Harry Bull.

Old Dutch Wind Mill in Golden Gate Park Calif. Pumping capacity 150.000 gals per minute

The Lick Monument San Francisco Calif. Erected in honor of the Celebrated Miner — Lick.

The Cliff House and rock Golden Gate Park Calif.
April 17th 1921

Seal Rocks from the Cliff House. Look close and see the seals.
April 17th 1921

The Naval Training Station Goat Island San Francisco Calif.

Looking to the right is the entrance of Golden Gate from the Pacific. April 17th 1921

GOLDEN GATE from Craigmont Hill Berkley Calif.

San Francisco from the bay off of the deck of the Piedmont a ferry boat that crosses the bay between Berkley and San Francisco

"PHOSTINT"
POST CARD
THE HOPI HOUSE is an exact reproduction of a Hopi Indian pueblo. It is an irregular stone structure plastered with adobe and rising three stories high.
Here are Hopi men, women and children. Some decorating and burning pottery, others spinning yarn and weaving squaw dresses and blankets, and perhaps some maiden weaving a basket.
Hopi land is the old province of Tusayan on the border of the Painted Desert.
Several rooms in the Hopi House are devoted to a rare and costly exhibit of Indian handicraft.
DEC 27
1922
ARIZ.
THIS SPACE MAY BE USED FOR MESSAGE
THIS SPACE FOR THE ADDRESS
Pawnee City
Nebraska
Best Wishes For A
Speedy Recovery

FIRST CAPITOL OF KANSAS
Mr and Mrs. F. W. Chase

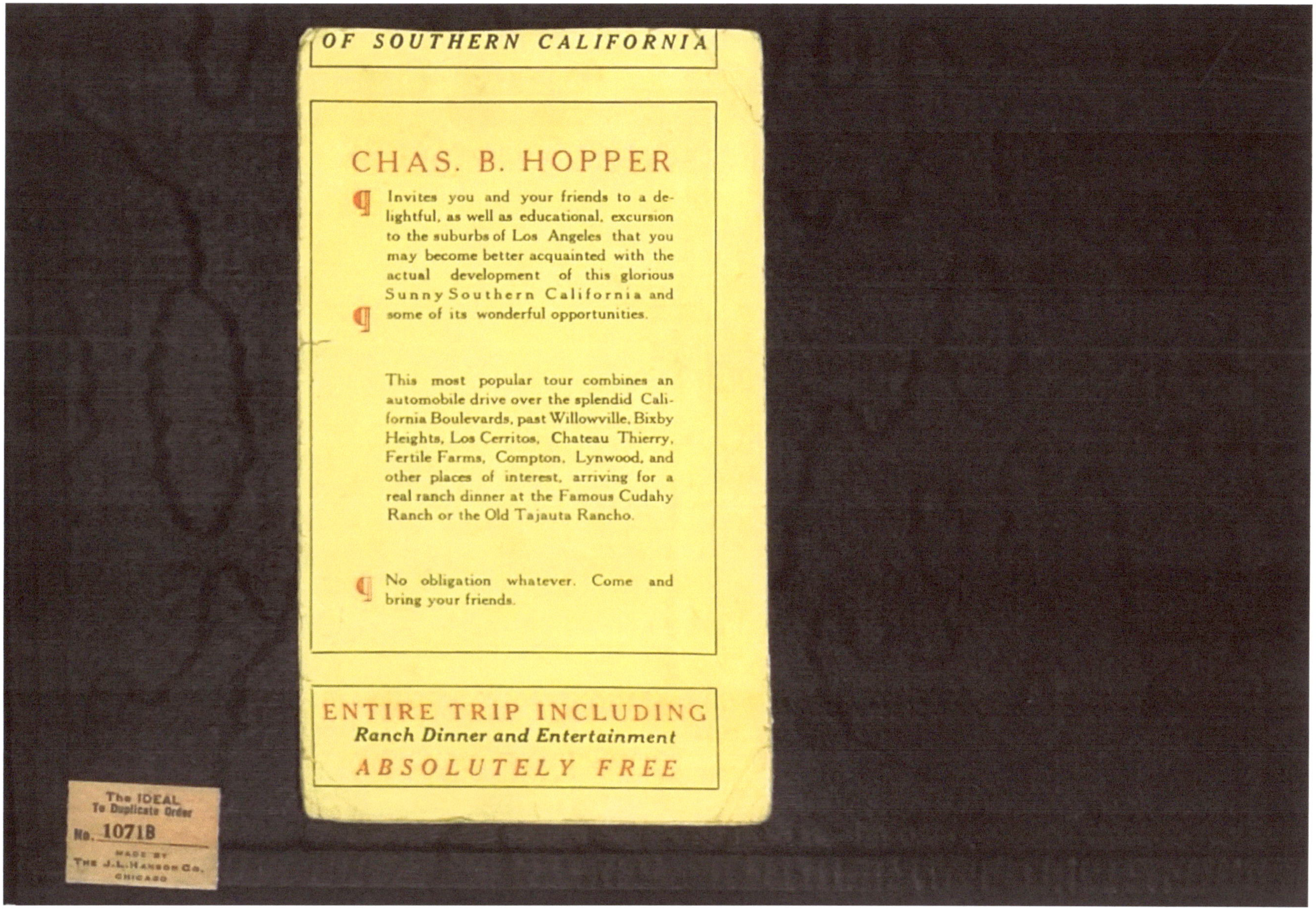
OF SOUTHERN CALIFORNIA
CHAS. B. HOPPER
Invites you and your friends to a delightful, as well as educational, excursion to the suburbs of Los Angeles that you may become better acquainted with the actual development of this glorious Sunny Southern California and some of its wonderful opportunities.
This most popular tour combines an automobile drive over the splendid California Boulevards, past Willowville, Bixby Heights, Los Cerritos, Chateau Thierry, Fertile Farms, Compton, Lynwood, and other places of interest, arriving for a real ranch dinner at the Famous Cudahy Ranch or the Old Tajauta Rancho.
No obligation whatever. Come and bring your friends.
ENTIRE TRIP INCLUDING
Ranch Dinner and Entertainment
ABSOLUTELY FREE
The IDEAL
To Duplicate Order
No. 1071B

CHASE FAMILY TIMELINE

June 16, 1853 — Fred Wilson Chase born in Whitingham, Windham County, Vermont, eighth in descent from Aquilla Chase, who crossed from England about 1636

June 9, 1856 — Anna Murdock born in Whitingham; she begins teaching country school at thirteen

December 24, 1874 — Fred and Anna marry on Christmas Eve; they start farming the next spring with one horse, two cows, a pair of yearling steers, and five hens

August 27, 1877 — Leon Wilson Chase, second of their nine children, born in Whitingham

Spring 1885 — After one last forenoon plowing around Vermont stones, Fred moves the family to Pawnee County, Nebraska, settling in a dug-out near Violet on the farm he names Hard Scrabble Ranch

About 1893 — The family tries Norton County, Kansas; eighteen months and two failed crops later ("the worst I ever experienced in farming"), they return to Pawnee County and rent

December 1899 — Fred makes the first $1,000 payment on the place known across eastern Nebraska as the Chase Farm; the family moves onto it in February 1901

1900s–1910s — Corn variety experiments, Holstein dairy records, and King Road Drag reports draw the attention of the State Agricultural College; Fred begins two decades of Farmers' Institute addresses, the first at a country church east of Beatrice

1914 — Leon receives from Iowa State College the first agricultural engineering degree ever granted

July 15, 1919 — The Nebraska Tractor Test Law takes effect; Leon, chairman of the university's agricultural engineering department, provided technical assistance in drafting it

November 1919 — The Chase Farm, its dairy herd, and equipment sell at auction, the land at $233.25 an acre; Fred and Anna retire to Pawnee City

March 26–29, 1920 — Early in their journey, Fred and Anna cross from Brownsville, Texas into Matamoros, Mexico

March 31–April 9, 1920 — Back in Lincoln, the first official Nebraska tractor test runs on a John Deere Waterloo Boy Model N

1920–1921 — The trip: twenty thousand miles, more than seventeen thousand by auto, through thirty-eight states, Mexico, and Canada

January 1922 — Fred recounts the journey to the student body of the Agricultural School at the State Farm in Lincoln

April 14, 1922 — Chancellor Samuel Avery presents Fred a certificate of honor at the Temple Theatre for the promotion of better agriculture and good roads, "the proudest day of my life"

December 24, 1924 — Fred and Anna celebrate their golden wedding with nearly all their children and grandchildren present

January 1, 1925 — Fred publishes his *A Brief Historical Sketch*

November 28, 1925 — Fred dies at his home in Pawnee City at 72, after a day's illness with heart trouble; he's buried in Pawnee City Cemetery, where the town still gathered at his grave each Memorial Day decades later

1982 — The University of Nebraska rededicates its Agricultural Engineering Hall as L.W. Chase Hall, the name it carries today

www.ingramcontent.com/pod-product-compliance
Lightning Source LLC
LaVergne TN
LVHW070818210926
848482LV00005B/290
* 9 7 8 1 9 4 5 8 3 4 3 5 6 *